DC Electrical Circuit
Laboratory Manual

James M. Fiore

Laboratory Manual for DC Electrical Circuit Analysis

by

James M. Fiore

Version 1.4.3, 28 October 2020

This **Laboratory Manual for DC Electrical Circuit Analysis, by James M. Fiore** is copyrighted under the terms of a Creative Commons license:

This work is freely redistributable for non-commercial use, share-alike with attribution

Published by James M. Fiore via dissidents

ISBN13: 978-1796777543

For more information or feedback, contact:

James Fiore, Professor
Electrical Engineering Technology
Mohawk Valley Community College
1101 Sherman Drive
Utica, NY 13501
jfiore@mvcc.edu

For the latest revisions, related titles, and links to low cost print versions, go to:
www.mvcc.edu/jfiore or www.dissidents.com

YouTube Channel: *Electronics with Professor Fiore*

Cover art, *Chapman's Contribution*, by the author

Introduction

This laboratory manual is intended for use in a DC electrical circuits course and is appropriate for two and four year electrical engineering technology curriculums. The manual contains sufficient exercises for a typical 15 week course using a two to three hour practicum period. The topics range from basic laboratory procedures and resistor identification through series-parallel circuits, mesh and nodal analysis, superposition, Thevenin's theorem, maximum power transfer theorem, and concludes with an introduction to capacitors and inductors. For equipment, each lab station should include a dual adjustable DC power supply and a quality DMM capable of reading DC voltage, current and resistance. A selection of standard value ¼ watt carbon film resistors ranging from a few ohms to a few mega ohms is required along with 10 kΩ and 100 kΩ potentiometers, 100 nF and 220 nF capacitors, and 1 mH and 10 mH inductors. A decade resistance box may also be useful.

Each exercise begins with an Objective and a Theory Overview. The Equipment List follows with space provided for serial numbers and measured values of components. Schematics are presented next along with the step-by-step procedure. All data tables are grouped together, typically with columns for the theoretical and experimental results, along with a column for the percent deviations between them. Finally, a group of appropriate questions are presented. For those with longer scheduled lab times, a useful addition is to simulate the circuit(s) with a SPICE-based tool such as Multisim, PSpice, TINA-TI, LTspice, or similar software, and compare those results to the theoretical and experimental results as well.

A companion laboratory manual for AC electrical circuits is also available. Other manuals in this series include Semiconductor Devices (diodes, bipolar transistors and FETs), Operational Amplifiers & Linear Integrated Circuits, Computer Programming with Python™ and Multisim™, and Embedded Controllers Using C and Arduino. Texts are available for DC and AC Electrical Circuit Analysis, Embedded Controllers, Op Amps & Linear Integrated Circuits, and Semiconductor Devices.

A Note from the Author

This work was borne out of the frustration of finding a lab manual that covered all of the appropriate material at sufficient depth while remaining readable and affordable for the students. It is used at Mohawk Valley Community College in Utica, NY, for our ABET accredited AAS program in Electrical Engineering Technology. I am indebted to my students, co-workers and the MVCC family for their support and encouragement of this project. While it would have been possible to seek a traditional publisher for this work, as a long-time supporter and contributor to freeware and shareware computer software, I have decided instead to release this using a Creative Commons non-commercial, share-alike license. I encourage others to make use of this manual for their own work and to build upon it. If you do add to this effort, I would appreciate a notification.

"Begin with the possible and move gradually towards the impossible"

-Robert Fripp

Table of Contents

1. The Electrical Laboratory 8
2. DC Sources and Metering 14
3. Resistor Color Code 18
4. Ohm's Law 24
5. Series DC Circuits 28
6. Parallel DC Circuits 32
7. Series-Parallel DC Circuits 36
8. Ladders and Bridges 40
9. Potentiometers and Rheostats . . . 44
10. Superposition Theorem 48
11. Thévenin's Theorem 52
12. Maximum Power Transfer 56
13. Nodal Analysis 60
14. Mesh Analysis 64
15. Capacitors and Inductors 68

Appendix A: Technical Report Guidelines . . 72
Appendix B: Example Technical Report . . 74
Appendix C: Creating Graphs Using a Spreadsheet 79
Appendix D: Using a Solderless Breadboard . . 81

1
The Electrical Laboratory

Objective

The laboratory emphasizes the practical, hands-on component of this course. It complements the theoretical material presented in lecture, and as such, is integral and indispensible to the mastery of the subject. There are several items of importance here including proper safety procedures, required tools, and laboratory reports. This exercise will finish with an examination of scientific and engineering notation, the standard form of representing and manipulating values.

Lab Safety and Tools

If proper procedures are followed, the electrical lab is a perfectly safe place in which to work. There are some basic rules: No food or drink is allowed in lab at any time. Liquids are of particular danger as they are ordinarily conductive. While the circuitry used in lab is normally of no shock hazard, some of the test equipment may have very high internal voltages that could be lethal (in excess of 10,000 volts). Spilling a bottle of water or soda onto such equipment could leave the experimenter in the receiving end of a severe shock. Similarly, items such as books and jackets should not be left on top of the test equipment as it could cause overheating.

Each lab bench is self contained. All test equipment is arrayed along the top shelf. Beneath this shelf at the back of the work area is a power strip. All test equipment for this bench should be plugged into this strip. None of this equipment should be plugged into any other strip. This strip is controlled by a single circuit breaker which also controls the bench light. In the event of an emergency, all test equipment may be powered off through this one switch. Further, the benches are controlled by dedicated circuit breakers in the front of the lab. Next to this main power panel is an A/B/C class fire extinguisher suitable for electrical fires. Located at the rear of the lab is a safety kit. This contains bandages, cleaning swaps and the like for small cuts and the like. For serious injury, the Security Office will be contacted.

A lab bench should always be left in a secure mode. This means that the power to each piece of test equipment should be turned off, the bench itself should be turned off, all AC and DC power and signal sources should be turned down to zero, and all other equipment and components properly stowed with lab stools pushed under the bench.

It is important to come prepared to lab. This includes the class text, the lab exercise for that day, class notebook, calculator, and hand tools. The tools include an electronic breadboard, test leads, wire strippers, and needle-nose pliers or hemostats. A small pencil soldering iron may also be useful. A basic DMM (digital multimeter) rounds out the list.

A typical breadboard or protoboard is shown below:

This particular unit features two main wiring sections with a common strip section down the center. Boards can be larger or smaller than this and may or may not have the mounting plate as shown. The connections are spaced 0.1 inch apart which is the standard spacing for many semiconductor chips. These are clustered in groups of five common terminals to allow multiple connections. The exception is the common strip which may have dozens of connection points. These are called *buses* and are designed for power and ground connections. Interconnections are normally made using small diameter solid hookup wire, usually AWG 22 or 24. Larger gauges may damage the board while smaller gauges do not always make good connections and are easy to break.

In the picture below, the color highlighted sections indicate common connection points. Note the long blue section which is a bus. This unit has four discrete buses available. When building circuits on a breadboard, it is important to keep the interconnecting wires short and the layout as neat as possible. This will aid both circuit functioning and ease of troubleshooting.

Examples of breadboard usage can be found in [Appendix D](). **Be sure to read it!**

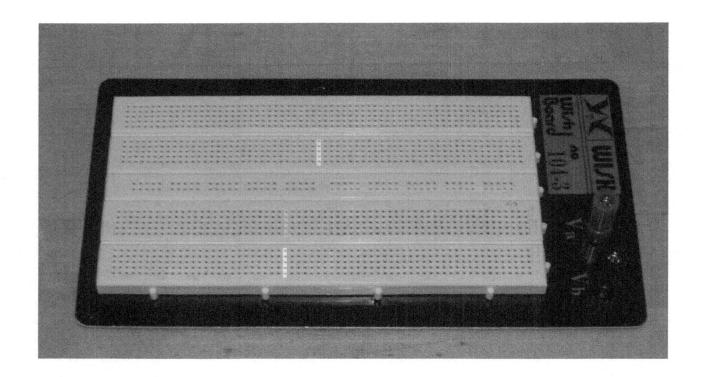

Laboratory Reports

Unless specified otherwise, all lab exercises require a non-formal laboratory report. Lab reports are individual endeavors not group work. The deadline for reports is one week after the exercise is performed. A letter grade is subtracted for the first half-week late and two letter grades are subtracted for the second half-week late. Reports are not acceptable beyond one week late. A basic report should include a statement of the Objective (i.e., those items under investigation), a Conclusion (what was found or verified), a Discussion (an explanation and analysis of the lab data which links the Objective to the Conclusion), Data Tables and Graphs, and finally, answers to any problems or questions posed in the exercise. Details of the structure of the report along with an example report may be found in Appendices A, B and C.

Scientific and Engineering Notation

Scientists and engineers often work with very large and very small numbers. The ordinary practice of using commas and leading zeroes proves to be very cumbersome in this situation. Scientific notation is more compact and less error prone method of representation. The number is split into two portions: a precision part (the mantissa) and a magnitude part (the exponent, being a power of ten). For example, the value 23,000 could be written as 23 times 10 to the 3rd power (that is, times one thousand). The exponent may be thought of in terms of how places the decimal point is moved to the left. Spelling this out is awkward, so a shorthand method is used where "times 10 to the X power" is replaced by the letter E (which stands for exponent). Thus, 23,000 could be written as 23E3. The value 45,000,000,000 would be written as 45E9. Note that it would also be possible to write this number as 4.5E10 or even 0.45E11. The only difference between scientific notation and engineering notation is that for engineering notation the exponent is always a multiple of three. Thus, 45E9 is proper engineering notation but 4.5E10 isn't. On most scientific calculators E is represented by either an "EE" or "EXP" button. The process of entering the value 45E9 would be depressing the keys 4 5 EE 9.

For fractional values, the exponent is negative and may be thought of in terms how many places the decimal point must be moved to the right. Thus, 0.00067 may be written as 0.67E−3 or 6.7E−4 or even 670E−6. Note that only the first and last of these three are acceptable as engineering notation.

Engineering notation goes one step further by using a set of prefixes to replace the multiples of three for the exponent. The prefixes are:

E12 = Tera (T)	E9 = Giga (G)	E6 = Mega (M)	E3 = kilo (k)
E−3 = milli (m)	E−6 = micro (µ)	E−9 = nano (n)	E−12 = pico (p)

Thus, 23,000 volts could be written as 23E3 volts or simply 23 kilovolts.

Besides being more compact, this notation is much simpler than the ordinary form when manipulating wide ranging values. When multiplying, simply multiply the precision portions and add the exponents. Similarly, when dividing, divide the precision portions and subtract the exponents. For example, 23,000 times 0.000003 may appear to be a complicated task. In engineering notation this is 23E3 times 3E−6. The result is 69E−3 (that is, 0.069). Given enough practice it will become second nature that kilo (E3) times micro (E−6) yields milli (E−3). This will facilitate lab estimates a great deal. Continuing, 42,000,000 divided by 0.002 is 42E6 divided by 2E−3, or 21E9 (the exponent is 6 minus a negative 3, or 9).

When adding or subtracting, first make sure that the exponents are the same (scaling if required) and then add or subtract the precision portions. For example, 2E3 plus 5E3 is 7E3. By comparison, 2E3 plus 5E6 is the same as 2E3 plus 5000E3, or 5002E3 (or 5.002E6).

Perform the following operations. Convert the following into scientific and engineering notation.

1. 1,500
2. 63,200,000
3. 0.0234
4. 0.000059
5. 170

Convert the following into normal longhand notation:

6. 1.23E3
7. 54.7E6
8. 2E−3
9. 27E−9
10. 4.39E7

Use the appropriate prefix for the following:

11. 4E6 volts
12. 5.1E3 feet
13. 3.3E−6 grams

Perform the following operations:

14. 5.2E6 + 1.7E6
15. 12E3 − 900
16. 1.7E3 • 2E6
17. 48E3 / 4E6
18. 20 / 4E3
19. 10 M • 2 k
20. 8 n / 2 m

2
DC Sources and Metering

Objective

The objective of this exercise is to become familiar with the operation and usage of basic DC electrical laboratory devices, namely DC power supplies and digital multimeters.

Theory Overview

The adjustable DC power supply is a mainstay of the electrical and electronics laboratory. It is indispensible in the prototyping of electronic circuits and extremely useful when examining the operation of DC systems. Of equal importance is the handheld digital multimeter or DMM. This device is designed to measure voltage, current, and resistance at a minimum, although some units may offer the ability to measure other parameters such as capacitance or transistor beta. Along with general familiarity of the operation of these devices, it is very important to keep in mind that no measurement device is perfect; their relative *accuracy*, *precision*, and *resolution* must be taken into account. *Accuracy* refers to how far a measurement is from that parameter's true value. *Precision* refers to the repeatability of the measurement, that is, the sort of variance (if any) that occurs when a parameter is measured several times. For a measurement to be valid, it must be both accurate and repeatable. Related to these characteristics is *resolution*. Resolution refers to the smallest change in measurement that may be discerned. For digital measurement devices this is ultimately limited by the number of significant digits available to display.

A typical DMM offers 3 ½ digits of resolution, the half-digit referring to a leading digit that is limited to zero or one. This is also known as a "2000 count display", meaning that it can show a minimum of 0000 and a maximum of 1999. The decimal point is "floating" in that it could appear anywhere in the sequence. Thus, these 2000 counts could range from 0.000 volts up to 1.999 volts, or 00.00 volts to 19.99 volts, or 000.0 volts to 199.9 volts, and so forth. With this sort of limitation in mind, it is very important to set the DMM to the lowest range that won't produce an overload in order to achieve the greatest accuracy.

A typical accuracy specification would be 1% of the reading plus two counts. "Reading" refers to the value displayed. If the 2 volt range was selected to read 1 volt (a measurement range of 0.000 to 1.999 for a 3 ½ digit meter), 1% would be 10 millivolts (0.01 volts). To this a further uncertainty of two counts (i.e., the finest digit) must be included. In this example, the finest digit is one millivolt (0.001 volts) so this adds another 2 millivolts for a total of 12 millivolts of potential inaccuracy. In other words, the value displayed by the meter could be as much as 12 millivolts higher or lower than the true value. For the 20 volt range the inaccuracy would be computed in like manner but notice that accuracy is lost because the lowest digit is larger (i.e., the "counts" represent a larger value). In this case, the counts portion jumps up to 20 mV for a total inaccuracy of 30 mV. Obviously, if a signal in the vicinity of, say, 1.3 volts was to be measured, greater accuracy will be obtained on the 2 volt scale than on either the 20 or 200 volt scales. In contrast, the 200 millivolt scale would produce an overload situation and cannot be used. Overloads are often indicated by either a flashing display or a readout of "OL". Finally, analog meters typically give a base accuracy in terms of a percentage of "full scale" (i.e., the selected scale or range) and not the signal itself, and obviously, there is no "counts" specification.

Equipment

(1) Adjustable DC power supply   model:_____ srn:_____
(1) Digital multimeter   model:_____ srn:_____
(1) Precision digital multimeter   model:_____ srn:_____

Procedure

1. Assume a general purpose 3 ½ digit DMM is being used. Its base accuracy is listed as 2% of reading plus 5 counts. Compute the inaccuracy caused by the scale and count factors and determine the total for a full scale reading. Record these values in Table 2.1.

2. Repeat step one for a precision 4 ½ digit DMM specified as 0.5% of reading plus 3 counts. Record the results in Table 2.2.

3. Set the adjustable power supply to 2.2 volts via its display. Use both the Coarse and Fine controls to get as close to 2.2 volts as possible. Record the displayed voltage in the first column of Table 2.3. Using the general purpose DMM set to the DC voltage function, set the range to 20 volts full scale. Measure the voltage at the output jacks of the power supply. Be sure to connect the DMM and power supply red lead to red lead, and black lead to black lead. Record the voltage registered by the DMM in the middle column of Table 2.3. Reset the DMM to the 200 volt scale, re-measure the voltage, and record in the final column

4. Repeat step three for the remaining voltages of Table 2.3.

5. Using the precision DMM, repeat steps three and four, recording the results in Table 2.4.

Data Tables

Scale	2% Reading	5 Counts	Total
200 mV			
20 V			

Table 2.1

Laboratory Manual for DC Electrical Circuit Analysis

Scale	.5% Reading	3 Counts	Total
200 mV			
2 V			

Table 2.2

Voltage	Power Supply	DMM 20V Scale	DMM 200V Scale
2.2			
5.0			
9.65			
15.0			

Table 2.3

Voltage	Power Supply	DMM 20V Scale	DMM 200V Scale
2.2			
5.0			
9.65			
15.0			

Table 2.4

Questions

1. For the general purpose DMM of Table 2.1, which contributes the larger share of inaccuracy; the full scale percentage or the count spec?

2. Bearing in mind that the power supply display is really just a very limited sort of digital volt meter, which voltages in Table 2.3 and 2.4 do you suspect to be the most accurately measured and why?

3. Assuming that the precision DMM used in Table 2.4 has a base accuracy spec of 0.1% plus 2 counts and is properly calibrated, what is the range of possible "true" voltages measured for 15.0 volts on the 20 volt scale?

3
Resistor Color Code

Objective
The objective of this exercise is to become familiar with the measurement of resistance values using a digital multimeter (DMM). A second objective is to learn the resistor color code.

Theory Overview
The resistor is perhaps the most fundamental of all electrical devices. Its fundamental attribute is the restriction of electrical current flow: The greater the resistance, the greater the restriction of current. Resistance is measured in ohms. The measurement of resistance in unpowered circuits may be performed with a digital multimeter. Like all components, resistors cannot be manufactured to perfection. That is, there will always be some variance of the true value of the component when compared to its nameplate or nominal value. For precision resistors, typically 1% tolerance or better, the nominal value is usually printed directly on the component. Normally, general purpose components, i.e. those worse than 1%, usually use a color code to indicate their value.

The resistor color code typically uses 4 color bands. The first two bands indicate the precision values (i.e. the mantissa) while the third band indicates the power of ten applied (i.e. the number of zeroes to add). The fourth band indicates the tolerance. It is possible to find resistors with five or six bands but they will not be examined in this exercise. Examples are shown below:

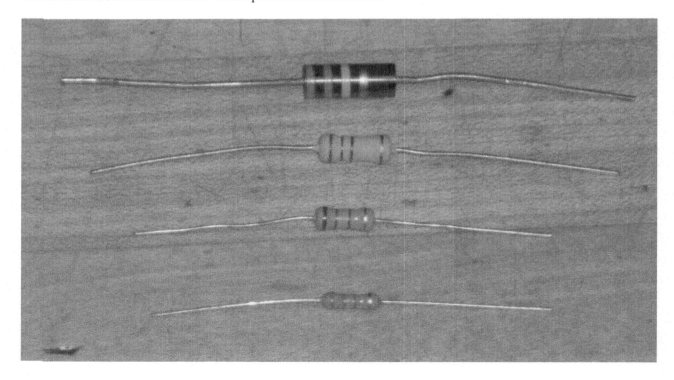

It is important to note that the physical size of the resistor indicates its power dissipation rating, not its ohmic value.

Each color in the code represents a numeral. It starts with black and finishes with white, going through the rainbow in between:

0 Black	1 Brown	2 Red	3 Orange	4 Yellow
5 Green	6 Blue	7 Violet	8 Gray	9 White

For the fourth, or tolerance, band:
5% Gold   10% Silver   20% None

For example, a resistor with the color code brown-red-orange-silver would correspond to 1 2 followed by 3 zeroes, or 12,000 ohms (more conveniently, 12 k ohms). It would have a tolerance of 10% of 12 k ohms or 1200 ohms. This means that the actual value of any particular resistor with this code could be anywhere between 12,000 − 1200=10,800, to 12,000 + 1200=13,200. That is, 10.8 k to 13.2 k ohms. Note, the IEC standard replaces the decimal point with the engineering prefix, thus 1.2 k is alternately written 1k2.

Similarly, a 470 k 5% resistor would have the color code yellow-violet-yellow-gold. To help remember the color code many mnemonics have been created using the first letter of the colors to create a sentence. One example is the *picnic mnemonic* **B**lack **B**ears **R**obbed **O**ur **Y**ummy **G**oodies **B**eating **V**arious **G**ray **W**olves.

Measurement of resistors with a DMM is a very straight forward process. Simply set the DMM to the resistance function and choose the first scale that is higher than the expected value. Clip the leads to the resistor and record the resulting value.

Equipment

(1) Digital multimeter   model:_____ srn:_____

Procedure

1. Given the nominal values and tolerances in Table 3.1, determine and record the corresponding color code bands.

2. Given the color codes in Table 3.2, determine and record the nominal value, tolerance and the minimum and maximum acceptable values.

3. Obtain a resistor equal to the first value listed in Table 3.3. Determine the minimum and maximum acceptable values based on the nominal value and tolerance. Record these values in Table 3.3. Using the DMM, measured the actual value of the resistor and record it in Table 3.3. Determine the deviation percentage of this component and record it in Table 3.3. The deviation percentage may be found via: Deviation = 100 * (measured − nominal)/nominal. *Circle the deviation if the resistor is out of tolerance.*

4. Repeat Step 3 for the remaining resistor in Table 3.3.

Laboratory Manual for DC Electrical Circuit Analysis

Data Tables

Value	Band 1	Band 2	Band 3	Band 4
27 @ 10%				
56 @ 10%				
180 @ 5%				
390 @ 10%				
680 @ 5%				
1.5 k @ 20%				
3.6 k @ 10%				
7.5 k @ 5%				
10 k @ 5%				
47 k @ 10%				
820 k @ 10%				
2.2 M @ 20 %				

Table 3.1

Colors	Nominal	Tolerance	Minimum	Maximum
red-red-black-silver				
blue-gray-black-gold				
brown-green-brown-gold				
orange-orange-brown-silver				
green-blue-brown –gold				
brown-red-red–silver				
red-violet-red–silver				
gray-red-red–gold				
brown-black-orange–gold				
orange-orange-orange–silver				
blue-gray-yellow–none				
green-black-green-silver				

Table 3.2

Value	Minimum	Maximum	Measured	Deviation
22 @ 10%				
68 @ 5%				
150 @ 5%				
330 @ 10%				
560 @ 5%				
1.2 k @ 5%				
2.7 k @ 10%				
8.2 k @ 5%				
10 k @ 5%				
33 k @ 10%				
680 k @ 10%				
5 M @ 20 %				

Table 3.3

Questions

1. What is the largest deviation in Table 3.3? Would it ever be possible to find a value that is outside the stated tolerance? Why or why not?

2. If Steps 3 and 4 were to be repeated with another batch of resistors, would the final two columns be identical to the original Table 3.3? Why or why not?

3. Do the measured values of Table 3.3 represent the exact values of the resistors tested? Why or why not?

4
Ohm's Law

Objective

This exercise examines Ohm's law, one of the fundamental laws governing electrical circuits. It states that voltage is equal to the product of current times resistance.

Theory Overview

Ohm's law is commonly written as $V = I * R$. That is, for a given current, an increase in resistance will result in a greater voltage. Alternately, for a given voltage, an increase in resistance will produce a decrease in current. As this is a first order linear equation, plotting current versus voltage for a fixed resistance will yield a straight line. The slope of this line is the conductance, and conductance is the reciprocal of resistance. Therefore, for a high resistance, the plot line will appear closer to the horizontal while a lower resistance will produce a more vertical plot line.

Equipment

(1) Adjustable DC power supply   model:_____ srn:_____
(1) Digital multimeter   model:_____ srn:_____
(1) 1 kΩ resistor   _____
(1) 6.8 kΩ resistor   _____
(1) 33 kΩ resistor   _____

Schematic

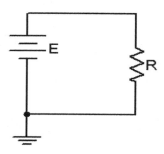

Figure 4.1

Procedure

1. Build the circuit of Figure 4.1 using the 1 kΩ resistor. Set the DMM to measure DC current and insert it in-line between the source and resistor. Set the source for zero volts. Measure and record the current in Table 4.1. Note that the theoretical current is 0 and any measured value other than 0 would produce an undefined percent deviation.

2. Setting E at 2 volts, determine the theoretical current based on Ohm's law and record this in Table 4.1. Measure the actual current, determine the deviation, and record these in Table 4.1. Note that Deviation = 100 * (measured − theory) / theory.

3. Repeat step 2 for the remaining source voltages in Table 4.1.

4. Remove the 1 kΩ and replace it with the 6.8 kΩ. Repeat steps 1 through 3 using Table 4.2.

5. Remove the 6.8 kΩ and replace it with the 33 kΩ. Repeat steps 1 through 3 using Table 4.3.

6. Using the measured currents from Tables 4.1, 4.2, and 4.3, create a plot of current versus voltage. Plot all three curves on the same graph. Voltage is the horizontal axis and current is the vertical axis.

Data Tables

E (volts)	I theory	I measured	Deviation
0	0		
2			
4			
6			
8			
10			
12			

Table 4.1 (1 kΩ)

E (volts)	I theory	I measured	Deviation
0	0		
2			
4			
6			
8			
10			
12			

Table 4.2 (6.8 kΩ)

E (volts)	I theory	I measured	Deviation
0	0		
2			
4			
6			
8			
10			
12			

Table 4.3 (33 kΩ)

Questions

1. Does Ohm's law appear to hold in this exercise?

2. Is there a linear relationship between current and voltage?

3. What is the relationship between the slope of the plot line and the circuit resistance?

5
Series DC Circuits

Objective

The focus of this exercise is an examination of basic series DC circuits with resistors. A key element is Kirchhoff's voltage law which states that the sum of voltage rises around a loop must equal the sum of the voltage drops. The voltage divider rule will also be investigated.

Theory Overview

A series circuit is defined by a single loop in which all components are arranged in daisy-chain fashion. The current is the same at all points in the loop and may be found by dividing the total voltage source by the total resistance. The voltage drops across any resistor may then be found by multiplying that current by the resistor value. Consequently, the voltage drops in a series circuit are directly proportional to the resistance. An alternate technique to find the voltage is the voltage divider rule. This states that the voltage across any resistor (or combination of resistors) is equal to the total voltage source times the ratio of the resistance of interest to the total resistance.

Equipment

(1) Adjustable DC power supply   model:_____ srn:_____
(1) Digital multimeter   model:_____ srn:_____
(1) 1 kΩ   _____
(1) 2.2 kΩ   _____
(1) 3.3 kΩ   _____
(1) 6.8 kΩ   _____

Schematics

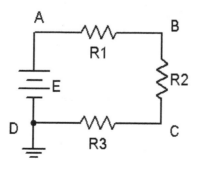

Figure 5.1

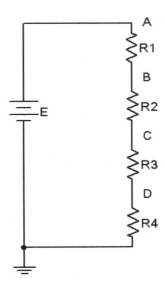

Figure 5.2

Procedure

1. Using the circuit of Figure 5.1 with R1 = 1 k, R2 = 2.2 k, R3 = 3.3 k, and E = 10 volts, determine the theoretical current and record it in Table 5.1. Construct the circuit. Set the DMM to read DC current and insert it in the circuit at point A. Remember, ammeters go in-line and require the circuit to be opened for proper measurement. The red lead should be placed closer to the positive source terminal. Record this current in Table 5.1. Repeat the current measurements at points B and C.

2. Using the theoretical current found in Step 1, apply Ohm's law to determine the expected voltage drops across R1, R2, and R3. Record these values in the Theory column of Table 5.2.

3. Set the DMM to measure DC voltage. Remember, unlike current, voltage is measured across components. Place the DMM probes across R1 and measure its voltage. Again, red lead should be placed closer to the positive source terminal. Record this value in Table 5.2. Repeat this process for the voltages across R2 and R3. Determine the percent deviation between theoretical and measured for each of the three resistor voltages and record these in the final column of Table 5.2.

4. Consider the circuit of Figure 5.2 with R1 = 1 k, R2 = 2.2 k, R3 = 3.3 k, R4 = 6.8 k and E = 20 volts. Using the voltage divider rule, determine the voltage drops across each of the four resistors and record the values in Table 5.3 under the Theory column. Note that the larger the resistor, the greater the voltage should be. Also determine the potentials V_{AC} and V_B, again using the voltage divider rule.

5. Construct the circuit of Figure 5.2 with R1 = 1 k, R2 = 2.2 k, R3 = 3.3 k, R4 = 6.8 k and E = 20 volts. Set the DMM to measure DC voltage. Place the DMM probes across R1 and measure its voltage. Record this value in Table 5.3. Also determine the deviation. Repeat this process for the remaining three resistors.

6. To find V_{AC}, place the red probe on point A and the black probe on point C. Similarly, to find V_B, place the red probe on point B and the black probe on ground. Record these values in Table 5.3 with deviations.

Simulation

7. Build the circuit of Figure 5.1 in a simulator. Using the virtual DMM as a voltmeter determine the voltages at nodes A, B and C, and compare these to the theoretical and measured values recorded in Table 5.2.

Data Tables

I Theory	I Point A	I Point B	I Point C

Table 5.1

Voltage	Theory	Measured	Deviation
R1			
R2			
R3			

Table 5.2

Voltage	Theory	Measured	Deviation
R1			
R2			
R3			
R4			
V_{AC}			
V_B			

Table 5.3

Questions

1. For the circuit of Figure 5.1, what is the expected current measurement at point D?

2. For the circuit of Figure 5.2, what are the expected current and voltage measurements at point D?

3. In Figure 5.2, R4 is approximately twice the size of R3 and about three times the size of R2. Would the voltages exhibit the same ratios? Why/why not? What about the currents through the resistors?

4. If a fifth resistor of 10 kΩ was added below R4 in Figure 5.2, how would this alter V_{AC} and V_B? Show work.

5. Is KVL satisfied in Tables 5.2 and 5.3?

6
Parallel DC Circuits

Objective

The focus of this exercise is an examination of basic parallel DC circuits with resistors. A key element is Kirchhoff's current law which states that the sum of currents entering a node must equal the sum of the currents exiting that node. The current divider rule will also be investigated.

Theory Overview

A parallel circuit is defined by the fact that all components share two common nodes. The voltage is the same across all components and will equal the applied source voltage. The total supplied current may be found by dividing the voltage source by the equivalent parallel resistance. It may also be found by summing the currents in all of the branches. The current through any resistor branch may be found by dividing the source voltage by the resistor value. Consequently, the currents in a parallel circuit are inversely proportional to the associated resistances. An alternate technique to find a particular current is the current divider rule. For a two resistor circuit this states that the current through one resistor is equal to the total current times the ratio of the other resistor to the total resistance.

Equipment

(1) Adjustable DC power supply   model:_____ srn:_____
(1) Digital multimeter   model:_____ srn:_____
(1) 1 kΩ   _____
(1) 2.2 kΩ   _____
(1) 3.3 kΩ   _____
(1) 6.8 kΩ   _____

Schematics

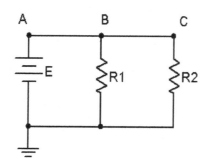

Figure 6.1

Laboratory Manual for DC Electrical Circuit Analysis

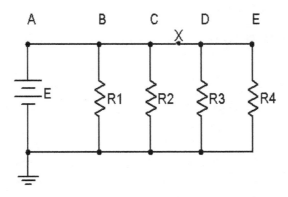

Figure 6.2

Procedure

1. Using the circuit of Figure 6.1 with R1 = 1 k, R2 = 2.2 k and E = 8 volts, determine the theoretical voltages at points A, B, and C with respect to ground. Record these values in Table 6.1. Construct the circuit. Set the DMM to read DC voltage and apply it to the circuit from point A to ground. The red lead should be placed at point A and the black lead should be connected to ground. Record this voltage in Table 6.1. Repeat the measurements at points B and C.

2. Apply Ohm's law to determine the expected currents through R1 and R2. Record these values in the Theory column of Table 6.2. Also determine and record the total current.

3. Set the DMM to measure DC current. Remember, current is measured at a single point and requires the meter to be inserted in-line. To measure the total supplied current place the DMM between points A and B. The red lead should be placed closer to the positive source terminal. Record this value in Table 6.2. Repeat this process for the currents through R1 and R2. Determine the percent deviation between theoretical and measured for each of the currents and record these in the final column of Table 6.2.

4. Crosscheck the theoretical results by computing the two resistor currents through the current divider rule. Record these in Table 6.3.

5. Consider the circuit of Figure 6.2 with R1 = 1 k, R2 = 2.2 k, R3 = 3.3 k, R4 = 6.8 k and E = 10 volts. Using the Ohm's law, determine the currents through each of the four resistors and record the values in Table 6.4 under the Theory column. Note that the larger the resistor, the smaller the current should be. Also determine and record the total supplied current and the current I_X. Note that this current should equal the sum of the currents through R3 and R4.

6. Construct the circuit of Figure 6.2 with R1 = 1 k, R2 = 2.2 k, R3 = 3.3 k, R4 = 6.8 k and E = 10 volts. Set the DMM to measure DC current. Place the DMM probes in-line with R1 and measure its current. Record this value in Table 6.4. Also determine the deviation. Repeat this process for the remaining three resistors. Also measure the total current supplied by the source by inserting the ammeter between points A and B.

Laboratory Manual for DC Electrical Circuit Analysis

7. To find I_X, insert the ammeter at point X with the black probe closer to R3. Record this value in Table 6.4 with deviation.

Simulation

8. Build the circuit of Figure 6.2 in a simulator. Using the virtual DMM as an ammeter determine the currents through the four resistors along with I_X, and compare these to the theoretical and measured values recorded in Table 6.4.

Data Tables

Voltage	Theory	Measured
V_A		
V_B		
V_C		

Table 6.1

Current	Theory	Measured	Deviation
R1			
R2			
Total			

Table 6.2

Current	CDR Theory
R1	
R2	
Total	

Table 6.3

Current	Theory	Measured	Deviation
R1			
R2			
R3			
R4			
Total			
I_x			

Table 6.4

Questions

1. For the circuit of Figure 6.1, what is the expected current entering the negative terminal of the source?

2. For the circuit of Figure 6.2, what is the expected current between points B and C?

3. In Figure 6.2, R4 is approximately twice the size of R3 and about three times the size of R2. Would the currents exhibit the same ratios? Why/why not?

4. If a fifth resistor of 10 kΩ was added to the right of R4 in Figure 6.2, how would this alter I_{Total} and I_X? Show work.

5. Is KCL satisfied in Tables 6.2 and 6.4?

7
Series-Parallel DC Circuits

Objective
This exercise will involve the analysis of basic series-parallel DC circuits with resistors. The use of simple series-only and parallel-only sub-circuits is examined as one technique to solve for desired currents and voltages.

Theory Overview
Simple series-parallel networks may be viewed as interconnected series and parallel sub-networks. Each of these sub-networks may be analyzed through basic series and parallel techniques such as the application of voltage divider and current divider rules along with Kirchhoff's voltage and current laws. It is important to identify the most simple series and parallel connections in order to jump to more complex interconnections.

Equipment

(1) Adjustable DC power supply   model:_____ srn:_____
(1) Digital multimeter   model:_____ srn:_____
(1) 1 kΩ   _____
(1) 2.2 kΩ   _____
(1) 3.3 kΩ   _____
(1) 4.7 kΩ   _____
(1) 6.8 kΩ   _____

Schematics

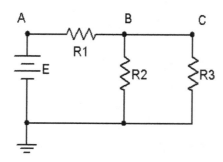

Figure 7.1

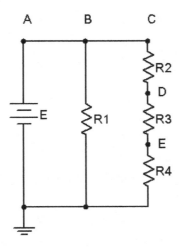

Figure 7.2

Procedure

1. Consider the circuit of Figure 7.1 with R1 = 1 k, R2 = 2.2 k, R3 = 4.7 k and E = 10 volts. R2 is in parallel with R3. This combination is in series with R1. Therefore, the R2, R3 pair may be treated as a single resistance to form a series loop with R1. Based on this observation, determine the theoretical voltages at points A, B, and C with respect to ground. Record these values in Table 7.1. Construct the circuit. Set the DMM to read DC voltage and apply it to the circuit from point A to ground. Record this voltage in Table 7.1. Repeat the measurements at points B and C, determine the deviations, and record the values in Table 7.1.

2. Applying KCL to the parallel sub-network, the current entering node B (i.e., the current through R1) should equal the sum of the currents flowing through R2 and R3. These currents may be determined through Ohm's law and/or the current divider rule. Compute these currents and record them in Table 7.2. Using the DMM as an ammeter, measure these three currents and record the values along with deviations in Table 7.2.

3. Consider the circuit of Figure 7.2. R2, R3 and R4 create a series sub-network. This sub-network is in parallel with R1. By observation then, the voltages at nodes A, B and C should be identical as in any parallel circuit of similar construction. Due to the series connection, the same current flows through R2, R3 and R4. Further, the voltages across R2, R3 and R4 should sum up to the voltage at node C, as in any similarly constructed series network. Finally, via KCL, the current exiting the source must equal the sum of the currents entering R1 and R2.

4. Build the circuit of Figure 7.2 with R1 = 3.3 k, R2 = 2.2 k, R3 = 4.7 k, R4 = 6.8 k and E = 20 volts. Using the series and parallel relations noted in Step 3, calculate the voltages at points B, C, D and E. Measure these potentials with the DMM, determine the deviations, and record the values in Table 7.3.

Laboratory Manual for DC Electrical Circuit Analysis

5. Calculate the currents leaving the source and flowing through R1 and R2. Record these values in Table 7.4. Using the DMM as an ammeter, measure those same currents, compute the deviations, and record the results in Table 7.4.

Simulation

6. Build the circuit of Figure 7.1 in a simulator. Using the virtual DMM as a voltmeter determine the voltages at nodes A, B and C, and compare these to the theoretical and measured values recorded in Table 7.1.

7. Build the circuit of Figure 7.2 in a simulator. Using the DC Operating Point simulation function, determine the voltages at nodes B, C, D and E, and compare these to the theoretical and measured values recorded in Table 7.3.

Data Tables

Voltage	Theory	Measured	Deviation
V_A			
V_B			
V_C			

Table 7.1

Current	Theory	Measured	Deviation
R_1			
R_2			
R_3			

Table 7.2

Voltage	Theory	Measured	Deviation
V_B			
V_C			
V_D			
V_E			

Table 7.3

Current	Theory	Measured	Deviation
Source			
R1			
R2			

Table 7.4

Questions

1. Are KVL and KCL satisfied in Tables 7.1 and 7.2?

2. Are KVL and KCL satisfied in Tables 7.3 and 7.4?

3. How would the voltages at A and B in Figure 7.1 change if a fourth resistor equal to 10 k was added in parallel with R3? What if this resistor was added in series with R3?

4. How would the currents through R1 and R2 in Figure 7.2 change if a fifth resistor equal to 10 k was added in series with R1? What if this resistor was added in parallel with R1?

8
Ladders and Bridges

Objective

The objective of this exercise is to continue the exploration of basic series-parallel DC circuits. The basic ladder network and bridge are examined. A key element here is the concept of loading, that is, the effect that a sub-circuit may have on a neighboring sub-circuit.

Theory Overview

Ladder networks are comprised of a series of alternating series and parallel connections. Each section effectively loads the prior section, meaning that the voltage and current of the prior section may change considerably if the loading section is removed. One possible technique for the solution of ladder networks is a series of cascading voltage dividers. Current dividers may also be used. In contrast, bridge networks typically make use of four elements arranged in dual series and parallel configuration. These are often used in measurement systems with the voltage of interest derived from the difference of two series sub-circuit voltages. As in the simpler series-parallel networks; KVL, KCL, the current divider rule and the voltage divider rule may be used in combination to analyze the sub-circuits.

Equipment

(1) Adjustable DC power supply   model:_____ srn:_____
(1) Digital multimeter   model:_____ srn:_____
(1) 1 kΩ _____
(1) 2.2 kΩ _____
(1) 3.3 kΩ _____
(1) 6.8 kΩ _____
(1) 10 kΩ _____
(1) 22 kΩ _____

Schematics

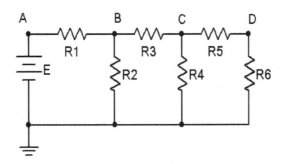

Figure 8.1

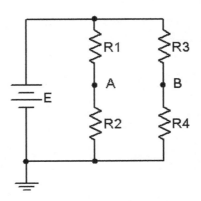

Figure 8.2

Procedure

1. Consider the circuit of Figure 8.1. R5 and R6 form a simple series connection. Together, they are in parallel with R4. Therefore the voltage across R4 must be the same as the sum of the voltages across R5 and R6. Similarly, the current entering node C from R3 must equal the sum of the currents flowing through R4 and R5. This three resistor combination is in series with R3 in much the same manner than R6 is in series with R5. These four resistors are in parallel with R2, and finally, these five resistors are in series with R1. Note that to find the voltage at node B the voltage divider rule may be used, *however*, it is important to note that VDR cannot be used in terms of R1 versus R2. Instead, R1 reacts against the entire series-parallel combination of R2 through R6. Similarly, R3 reacts against the combination of R4, R5 and R6. That is to say R5 and R6 *load* R4, and R3 through R6 *load* R2. Because of this process note that V_D must be less than V_C, which must be less than V_B, which must be less than V_A. Thus the circuit may be viewed as a sequence of loaded voltage dividers.

2. Construct the circuit of Figure 8.1 using R1 = 1 k, R2 = 2.2 k, R3 = 3.3 k, R4 = 6.8 k, R5 = 10 k, R6 = 22 k and E = 20 volts. Based on the observations of Step 1, determine the theoretical voltages at nodes A, B, C and D, and record them in Table 8.1. Measure the potentials with a DMM, compute the deviations and record the results in Table 8.1.

3. Based on the theoretical voltages found in Table 8.1, determine the currents through R1, R2, R4 and R6. Record these values in Table 8.2. Measure the currents with a DMM, compute the deviations and record the results in Table 8.2.

4. Consider the circuit of Figure 8.2. In this bridge network, the voltage of interest is V_{AB}. This may be directly computed from $V_A - V_B$. Assemble the circuit using R1 = 1 k, R2 = 2.2 k, R3 = 10 k, R4 = 6.8 k and E = 10 volts. Determine the theoretical values for V_A, V_B and V_{AB} and record them in Table 8.3. Note that the voltage divider rule is very effective here as the R1 R2 branch and the R3 R4 branch are in parallel and therefore both "see" the source voltage.

5. Use the DMM to measure the potentials at A and B with respect to ground, the red lead going to the point of interest and the black lead going to ground. To measure the voltage from A to B, the red lead is connected to point A while the black is connected to point B. Record these potentials in Table 8.3. Determine the deviations and record these in Table 8.3.

Data Tables

Voltage	Theory	Measured	Deviation
V_A			
V_B			
V_C			
V_D			

Table 8.1

Current	Theory	Measured	Deviation
R_1			
R_2			
R_4			
R_6			

Table 8.2

Voltage	Theory	Measured	Deviation
V_A			
V_B			
V_{AB}			

Table 8.3

Questions

1. In Figure 8.1, if another pair of resistors was added across R6, would V_D go up, down, or stay the same? Why?

2. In Figure 8.1, if R4 was accidentally opened would this change the potentials at B, C and D? Why or why not?

3. If the DMM leads are reversed in Step 5, what happens to the measurements in Table 8.3?

4. Suppose that R3 and R4 are accidentally swapped in Figure 8.2. What is the new V_{AB}?

9
Potentiometers and Rheostats

Objective
The objective of this exercise is to examine the practical workings of adjustable resistances, namely the potentiometer and rheostat. Their usage in adjustable voltage and current control will be investigated.

Theory Overview
A potentiometer is a three terminal resistive device. The outer terminals present a constant resistance which is the nominal value of the device. A third terminal, called the wiper arm, is in essence a contact point that can be moved along the resistance. Thus, the resistance seen from one outer terminal to the wiper plus the resistance from the wiper to the other outer terminal will always equal the nominal resistance of the device. This three terminal configuration is used typically to adjust voltage via the voltage divider rule, hence the name potentiometer, or *pot* for short. While the resistance change is often linear with rotation (i.e., rotating the shaft 50% yields 50% resistance), other schemes, called *tapers*, are also found. One common non-linear taper is the logarithmic taper. It is important to note that linearity can be compromised (sometimes on purpose) if the resistance loading the potentiometer is not significantly larger in value than the potentiometer itself.

If only a single outer terminal and the wiper are used, the device is merely an adjustable resistor and is referred to as a rheostat. These may be placed in-line with a load to control the load current, the greater the resistance, the smaller the current.

Equipment

(1) Adjustable DC power supply   model:_____ srn:_____
(1) Digital multimeter   model:_____ srn:_____
(1) 10 kΩ potentiometer
(1) 100 kΩ potentiometer
(1) 1 kΩ _____
(1) 4.7 kΩ _____
(1) 47 kΩ _____

Schematics

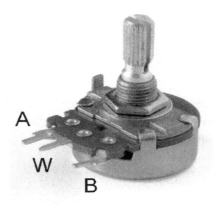

Figure 9.1

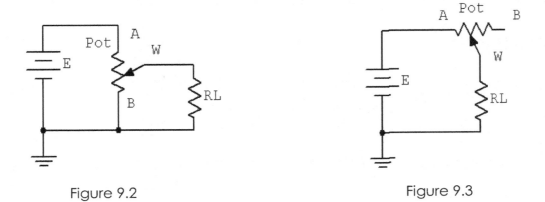

Figure 9.2   Figure 9.3

Procedure

1. A typical potentiometer is shown in Figure 9.1. Using a 10 k pot, first rotate the knob fully counter-clockwise and using the DMM, measure the resistance from terminal A to the wiper arm, W. Then measure the value from the wiper arm to terminal B. Record these values in Table 9.1. Add the two readings, placing the result in the final column.

2. Rotate the knob 1/4 turn clockwise and repeat the measurements of step 1. Repeat this process for the remaining knob positions in Table 9.1. Note that the results of the final column should all equal the nominal value of the potentiometer.

3. Construct the circuit of Figure 9.2 using E = 10 volts, a 10 k potentiometer and leave R_L open. Rotate the knob fully counter-clockwise and measure the voltage from the wiper to ground. Record this value in Table 9.2. Continue taking and recording voltages as the knob is rotated to the other four positions in Table 9.2.

Laboratory Manual for DC Electrical Circuit Analysis

4. Set R_L to 47 k and repeat step 3.

5. Set R_L to 4.7 k and repeat step 3.

6. Set R_L to 1 k and repeat step 3.

7. Using a linear grid, plot the voltages of Table 9.2 versus position. Note that there will be four curves created, one for each load, but place them on a single graph. Note how the variance of the load affects the linearity and control of the voltage.

8. Construct the circuit of Figure 9.3 using E = 10 volts, a 100 k potentiometer and R_L = 1 k. Rotate the knob fully counter-clockwise and measure the current through the load. Record this value in Table 9.3. Repeat this process for the remaining knob positions in Table 9.3.

9. Replace the load resistor with a 4.7 k and repeat step 8.

Data Tables

Position	R_{AW}	R_{WB}	$R_{AW} + R_{WB}$
Fully CCW			
1/4			
1/2			
3/4			
Fully CW			

Table 9.1

Position	V_{WB} Open	V_{WB} 47k	V_{WB} 4.7k	V_{WB} 1k
Fully CCW				
1/4				
1/2				
3/4				
Fully CW				

Table 9.2

Position	I_L 1k	I_L 4.7k
Fully CCW		
1/4		
1/2		
3/4		
Fully CW		

Table 9.3

Questions

1. In Table 9.1, does the total resistance always equal the nominal resistance of the potentiometer?

2. If the potentiometer used for Table 9.1 had a logarithmic taper, how would the values change?

3. In Table 9.2, is the load voltage always directly proportional to the knob position? Is the progression always linear?

4. Explain the variation of the four curves plotted in step 7.

5. In the final circuit, is the load current always proportional to the knob position? If the load was much smaller, say just a few hundred ohms, would the minimum and maximum currents be much different from those in Table 9.3?

6. How could the circuit of Figure 9.3 be modified so that the maximum current could be set to a value higher than that achieved by the supply and load resistor alone?

10
Superposition Theorem

Objective

The objective of this exercise is to investigate the application of the superposition theorem to multiple DC source circuits in terms of both voltage and current measurements. Power calculations will also be examined.

Theory Overview

The superposition theorem states that in a linear bilateral multi-source DC circuit, the current through or voltage across any particular element may be determined by considering the contribution of each source independently, with the remaining sources replaced with their internal resistance. The contributions are then summed, paying attention to polarities, to find the total value. Superposition cannot in general be applied to non-linear circuits or to non-linear functions such as power.

Equipment

(1) Adjustable dual DC power supply   model:_____ srn:_____
(1) Digital multimeter   model:_____ srn:_____
(1) 4.7 kΩ   _____
(1) 6.8 kΩ   _____
(1) 10 kΩ   _____
(1) 22 kΩ   _____
(1) 33 kΩ   _____

Schematics

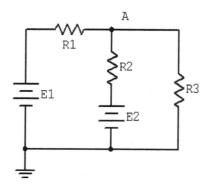

Figure 10.1

48   Laboratory Manual for DC Electrical Circuit Analysis

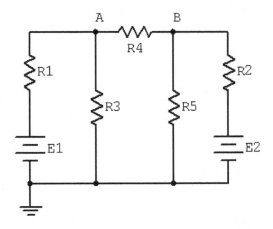

Figure 10.2

Procedure

Voltage Application

1. Consider the dual supply circuit of Figure 10.1 using E1 = 10 volts, E2 = 15 volts, R1 = 4.7 k, R2 = 6.8 k and R3 = 10 k. To find the voltage from node A to ground, superposition may be used. Each source is considered by itself. First consider source E1 by assuming that E2 is replaced with its internal resistance (a short). Determine the voltage at node A using standard series-parallel techniques and record it in Table 10.1. Make sure to indicate the polarity. Repeat the process using E2 while shorting E1. Finally, sum these two voltages and record in Table 10.1.

2. To verify the superposition theorem, the process may be implemented directly by measuring the contributions. Build the circuit of Figure 10.1 with the values specified in step 1, however, replace E2 with a short. Do **not** simply place a shorting wire across source E2! This will overload the power supply.

3. Measure the voltage at node A and record in Table 10.1. Be sure to note the polarity.

4. Remove the shorting wire and insert source E2. Also, replace source E1 with a short. Measure the voltage at node A and record in Table 10.1. Be sure to note the polarity.

5. Remove the shorting wire and re-insert source E1. Both sources should now be in the circuit. Measure the voltage at node A and record in Table 10.1. Be sure to note the polarity. Determine and record the deviations between theory and experimental results.

Current and Power Application

6. Consider the dual supply circuit of Figure 10.2 using E1 = 10 volts, E2 = 15 volts, R1 = 4.7 k, R2 = 6.8 k, R3 = 10 k, R4 = 22 k and R5 = 33 k. To find the current through R4 flowing from node A

to B, superposition may be used. Each source is again treated independently with the remaining sources replaced with their internal resistances. Calculate the current through R4 first considering E1 and then considering E2. Sum these results and record the three values in Table 10.2.

7. Assemble the circuit of Figure 10.2 using the values specified. Replace source E2 with a short and measure the current through R4. Be sure to note the direction of flow and record the result in Table 10.2.

8. Replace the short with source E2 and swap source E1 with a short. Measure the current through R4. Be sure to note the direction of flow and record the result in Table 10.2.

9. Remove the shorting wire and re-insert source E1. Both sources should now be in the circuit. Measure the current through R4 and record in Table 10.2. Be sure to note the direction. Determine and record the deviations between theory and experimental results.

10. Power is not a linear function as it is proportional to the square of either voltage or current. Consequently, superposition should not yield an accurate result when applied directly to power. Based on the measured currents in Table 10.2, calculate the power in R4 using E1-only and E2-only and record the values in Table 10.3. Adding these two powers yields the power as predicted by superposition. Determine this value and record it in Table 10.3. The true power in R4 may be determined from the total measured current flowing through it. Using the experimental current measured when both E1 and E2 were active (Table 10.2), determine the power in R4 and record it in Table 10.3.

Simulation

11. Build the circuit of Figure 10.2 in a simulator. Using the virtual DMM as an ammeter, determine the current through resistor R4 and compare it to the theoretical and measured values recorded in Table 10.2.

Data Tables

Source	V_A Theory	V_A Experimental	Deviation
E1 Only			
E2 Only			
E1 and E2			

Table 10.1

Source	I_{R4} Theory	I_{R4} Experimental	Deviation
E1 Only			
E2 Only			
E1 and E2			

Table 10.2

Source	P_{R4}
E1 Only	
E2 Only	
E1 + E2	
E1 and E2	

Table 10.3

Questions

1. Based on the results of Tables 10.1, 10.2 and 10.3, can superposition be applied successfully to voltage, current and power levels in a DC circuit?

2. If one of the sources in Figure 10.1 had been inserted with the opposite polarity, would there be a significant change in the resulting voltage at node A? Could both the magnitude and polarity change?

3. If both of the sources in Figure 10.1 had been inserted with the opposite polarity, would there be a significant change in the resulting voltage at node A? Could both the magnitude and polarity change?

4. Why is it important to note the polarities of the measured voltages and currents?

11
Thévenin's Theorem

Objective

The objective of this exercise is to examine the use of Thévenin's theorem to create simpler versions of DC circuits as an aide to analysis. Multiple methods of experimentally obtaining the Thévenin resistance will be explored.

Theory Overview

Thévenin's theorem for DC circuits states that any single port (i.e., two terminal points) linear network may be replaced by a single voltage source with an appropriate internal resistance. The Thévenin equivalent will produce the same load current and voltage as the original circuit to any load. Consequently, if many different loads or sub-circuits are under consideration, using a Thévenin equivalent may prove to be a quicker analysis route than "reinventing the wheel" each time.

The Thévenin voltage is found by determining the open circuit output voltage. The Thévenin resistance is found by replacing any DC sources with their internal resistances and determining the resulting combined resistance as seen from the two terminals using standard series-parallel analysis techniques. In the laboratory, the Thévenin resistance may be found using an ohmmeter (again, replacing the sources with their internal resistances) or by using the matched load technique. The matched load technique involves replacing the load with a variable resistance and then adjusting it until the load voltage is precisely one half of the unloaded voltage. This would imply that the other half of the voltage must be dropped across the equivalent Thévenin resistance, and as the Thévenin circuit is a simple series loop then the two resistances must be equal as they have identical currents and voltages.

Equipment

(1) Adjustable DC power supply   model:_____ srn:_____
(1) Digital multimeter   model:_____ srn:_____
(1) 2.2 kΩ   _____
(1) 3.3 kΩ   _____
(1) 4.7 kΩ   _____
(1) 6.8 kΩ   _____
(1) 8.2 kΩ   _____
(1) 10 k potentiometer or resistance decade box

Schematics

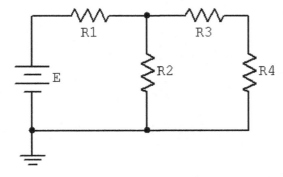

Figure 11.1

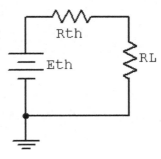

Figure 11.2

Procedure

1. Consider the circuit of Figure 11.1 using E = 10 volts, R1 = 3.3 k, R2 = 6.8 k, R3 = 4.7 k and R4 (R_{Load}) = 8.2 k. This circuit may be analyzed using standard series-parallel techniques. Determine the voltage across the load, R4, and record it in Table 11.1. Repeat the process using 2.2 k for R4.

2. Build the circuit of Figure 11.1 using the values specified in step one, with R_{Load} = 8.2 k. Measure the load voltage and record it in Table 11.1. Repeat this with a 2.2 k load resistance. Determine and record the deviations. Do not deconstruct the circuit.

3. Determine the theoretical Thévenin voltage of the circuit of Figure 11.1 by finding the open circuit output voltage. That is, replace the load with an open and calculate the voltage produced between the two open terminals. Record this voltage in Table 11.2.

4. To calculate the theoretical Thévenin resistance, first remove the load and then replace the source with its internal resistance (ideally, a short). Finally, determine the combination series-parallel resistance as seen from the where the load used to be. Record this resistance in Table 11.2.

5. The experimental Thévenin voltage maybe determined by measuring the open circuit output voltage. Simply remove the load from the circuit of step one and then replace it with a voltmeter. Record this value in Table 11.2.

6. There are two methods to measure the experimental Thévenin resistance. For the first method, using the circuit of step one, replace the source with a short. Then replace the load with the ohmmeter. The Thévenin resistance may now be measured directly. Record this value in Table 11.2.

7. In powered circuits, ohmmeters are not effective while power is applied. An alternate method relies on measuring the effect of the load resistance. Return the voltage source to the circuit, replacing the short from step six. For the load, insert either the decade box or the potentiometer. Adjust this device

until the load voltage is half of the open circuit voltage measured in step five and record in Table 11.2 under "Method 2". At this point, the load and the Thévenin resistance form a simple series loop as seen in Figure 11.2. This means that they "see" the same current. If the load exhibits one half of the Thévenin voltage then the other half must be dropped across the Thévenin resistance, in other words $V_{RL} = V_{RTH}$. Consequently, the resistances have the same voltage and current, and therefore must have the same resistance according to Ohm's law.

8. Consider the Thévenin equivalent of Figure 11.2 using the theoretical E_{TH} and R_{TH} from Table 11.2 along with 8.2 k for the load (RL). Calculate the load voltage and record it in Table 11.3. Repeat the process for a 2.2 k load.

9. Build the circuit of Figure 11.2 using the measured E_{TH} and R_{TH} from Table 11.2 along with 8.2 k for the load (RL). Measure the load voltage and record it in Table 11.3. Also determine and record the deviation.

10. Repeat step nine using a 2.2 k load.

Data Tables

Original Circuit

R_4 (Load)	V_{Load} Theory	V_{Load} Experimental	Deviation
8.2 k			
2.2 k			

Table 11.1

Thévenized Circuit

	Theory	Experimental
E_{TH}		
R_{TH}		
R_{TH} Method 2	X	

Table 11.2

R₄ (Load)	V_Load Theory	V_Load Experimental	Deviation
8.2 k			
2.2 k			

Table 11.3

Questions

1. Do the load voltages for the original and Thévenized circuits match for both loads? Is it logical that this could be extended to any arbitrary load resistance value?

2. Assuming several loads were under consideration, which is faster, analyzing each load with the original circuit of Figure 11.1 or analyzing each load with the Thévenin equivalent of Figure 11.2?

3. How would the Thévenin equivalent computations change if the original circuit contained more than one voltage source?

12

Maximum Power Transfer

Objective

The objective of this exercise is to determine the conditions under which a load will produce maximum power. Further, the variance of load power and system efficiency will be examined graphically.

Theory Overview

In order to achieve the maximum load power in a DC circuit, the load resistance must equal the driving resistance, that is, the internal resistance of the source. Any load resistance value above or below this will produce a smaller load power. System efficiency (η) is 50% at the maximum power case. This is because the load and the internal resistance form a basic series loop, and as they have the same value, they must exhibit equal currents and voltages, and hence equal powers. As the load increases in resistance beyond the maximizing value the load voltage will rise, however, the load current will drop by a greater amount yielding a lower load power. Although this is not the maximum load power, this will represent a larger percentage of total power produced, and thus a greater efficiency (the ratio of load power to total power).

Equipment

(1) Adjustable DC power supply   model:_____ srn:_____
(1) Digital multimeter   model:_____ srn:_____
(1) Resistance decade box
(1) 3.3 kΩ _____

Schematics

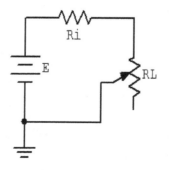

Figure 12.1

Procedure

1. Consider the simple the series circuit of Figure 12.1 using E = 10 volts and Ri = 3.3 k. Ri forms a simple voltage divider with RL. The power in the load is V_L^2/R_L and the total circuit power is $E^2/(R_i+R_L)$. The larger the value of RL, the greater the load voltage, however, this does not mean that very large values of RL will produce maximum load power due to the division by RL. That is, at some point V_L^2 will grow more slowly than RL itself. This crossover point should occur when RL is equal to Ri. Further, note that as RL increases, total circuit power decreases due to increasing total resistance. This should lead to an increase in efficiency. An alternate way of looking at the efficiency question is to note that as RL increases, circuit current decreases. As power is directly proportional to the square of current, as RL increases the power in Ri must decrease leaving a larger percentage of total power going to RL.

2. Using RL = 30, compute the expected values for load voltage, load power, total power and efficiency, and record them in Table 12.1. Repeat for the remaining RL values in the Table. For the middle entry labeled Actual, insert the measured value of the 3.3 k used for Ri.

3. Build the circuit of Figure 12.1 using E = 10 volts and Ri = 3.3 k. Use the decade box for RL and set it to 30 ohms. Measure the load voltage and record it in Table 12.2. Calculate the load power, total power and efficiency, and record these values in Table 12.2. Repeat for the remaining resistor values in the table.

4. Create two plots of the load power versus the load resistance value using the data from the two tables, one for theoretical, one for experimental. For best results make sure that the horizontal axis (RL) uses a log scaling instead of linear.

5. Create two plots of the efficiency versus the load resistance value using the data from the two tables, one for theoretical, one for experimental. For best results make sure that the horizontal axis (RL) uses a log scaling instead of linear.

Data Tables

R_L	V_L	P_L	P_T	η
30				
150				
500				
1 k				
2.5 k				
Actual=				
4 k				
10 k				
25 k				
70 k				
300 k				

Table 12.1

R_L	V_L	P_L	P_T	η
30				
150				
500				
1 k				
2.5 k				
Actual=				
4 k				
10 k				
25 k				
70 k				
300 k				

Table 12.2

Questions

1. At what point does maximum load power occur?

2. At what point does maximum total power occur?

3. At what point does maximum efficiency occur?

4. Is it safe to assume that generation of maximum load power is always a desired goal? Why/why not?

13
Nodal Analysis

Objective

The study of nodal analysis is the objective of this exercise, specifically its usage in multi-source DC circuits. Its application to finding circuit currents and voltages will be investigated.

Theory Overview

Multi-source DC circuits may be analyzed using a node voltage technique. The process involves identifying all of the circuit nodes, a node being a point where various branch currents combine. A reference node, usually ground, is included. Kirchhoff's current law is then applied to each node. Consequently a set of simultaneous equations are created with an unknown voltage for each node with the exception of the reference. In other words, a circuit with a total of five nodes including the reference will yield four unknown node voltages and four equations. Once the node voltages are determined, various branch currents and component voltages may be derived.

Equipment

(1) Adjustable DC power supply   model:_____ srn:_____
(1) Digital multimeter   model:_____ srn:_____
(1) 4.7 kΩ   _____
(1) 6.8 kΩ   _____
(1) 10 kΩ   _____
(1) 22 kΩ   _____
(1) 33 kΩ   _____

Schematics

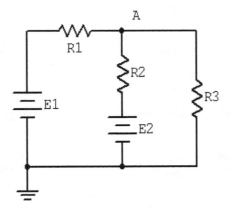

Figure 13.1

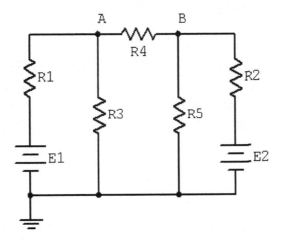

Figure 13.2

Procedure

1. Consider the dual supply circuit of Figure 13.1 using E1 = 10 volts, E2 = 15 volts, R1 = 4.7 k, R2 = 6.8 k and R3 = 10 k. To find the voltage from node A to ground, nodal analysis may be applied. In this circuit note that there is only one node and therefore only one equation with one unknown is needed. Once this potential is found, all other circuit currents and voltages may be found by applying Ohm's law and/or KVL and KCL.

2. Write the node equation for the circuit of Figure 13.1 and solve for node voltage A. Also, determine the current through R3. Record these values in Table 13.1.

3. Construct the circuit of Figure 13.1 using the values specified in step one. Measure the voltage from node A to ground along with the current though R3. Record these values in Table 13.1. Also determine and record the deviations.

4. Consider the dual supply circuit of Figure 13.2 using E1 = 10 volts, E2 = 15 volts, R1 = 4.7 k, R2 = 6.8 k, R3 = 10 k, R4 = 22 k and R5 = 33 k. Applying nodal analysis to this circuit yields two equations with two unknowns, namely node voltages A and B. Again, once these potentials are found, any other circuit current or voltage may be determined by applying Ohm's law and/or KVL and KCL.

5. Write the node equations for the circuit of Figure 13.2 and solve for node voltage A, node voltage B and the potential from A to B. Also, determine the current through R4. Record these values in Table 13.2.

6. Construct the circuit of Figure 13.2 using the values specified in step four. Measure the voltages from node A to ground, node B to ground and from node A to B, along with the current though R4. Record these values in Table 13.2. Also determine and record the deviations.

Simulation

7. Build the circuit of Figure 13.2 in a simulator. Using the DC Operating Point simulation function, determine the voltages at nodes A and B, and compare these to the theoretical and measured values recorded in Table 13.2.

Data Tables

Parameter	Theory	Experimental	Deviation
V_A			
I_{R3}			

Table 13.1

Parameter	Theory	Experimental	Deviation
V_A			
V_B			
V_{AB}			
I_{R4}			

Table 13.2

Questions

1. Do the polarities of the sources in Figure 13.1 matter as to the resulting voltages? Will the magnitudes of the voltages be the same if one or both sources have an inverted polarity?

2. In both circuits of this exercise the negative terminals of the sources are connected to ground. Is this a requirement for nodal analysis? What would happen to the node voltages if the positions of E1 and R1 in Figure 13.1 were swapped?

3. The circuits of Figures 13.1 and 13.2 had been analyzed previously in the superposition theorem exercise. How do the results of this exercise compare to the earlier results? Should the resulting currents and voltages be identical? If not, what sort of things might affect the outcome?

14

Mesh Analysis

Objective

The study of mesh analysis is the objective of this exercise, specifically its usage in multi-source DC circuits. Its application to finding circuit currents and voltages will be investigated.

Theory Overview

Multi-source DC circuits may be analyzed using a mesh current technique. The process involves identifying a minimum number of small loops such that every component exists in at least one loop. Kirchhoff's voltage law is then applied to each loop. The loop currents are referred to as mesh currents as each current interlocks or meshes with the surrounding loop currents. As a result there will be a set of simultaneous equations created, an unknown mesh current for each loop. Once the mesh currents are determined, various branch currents and component voltages may be derived.

Equipment

(1) Adjustable DC power supply   model:_____ srn:_____
(1) Digital multimeter   model:_____ srn:_____
(1) 4.7 kΩ _____
(1) 6.8 kΩ _____
(1) 10 kΩ _____
(1) 22 kΩ _____
(1) 33 kΩ _____

Schematics

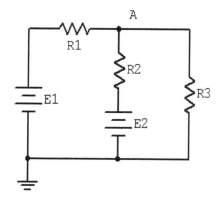

Figure 14.1

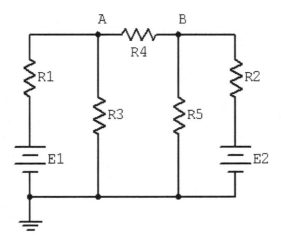

Figure 14.2

Procedure

1. Consider the dual supply circuit of Figure 14.1 using E1 = 10 volts, E2 = 15 volts, R1 = 4.7 k, R2 = 6.8 k and R3 = 10 k. To find the voltage from node A to ground, mesh analysis may be used. This circuit may be described via two mesh currents, loop one formed with E1, R1, R2 and E2, and loop two formed with E2, R2 and R3. Note that these mesh currents are the currents flowing through R1 and R3 respectively.

2. Using KVL, write the loop expressions for these two loops and then solve to find the mesh currents. Note that the third branch current (that of R2) is the combination of the mesh currents and that the voltage at node A can be determined using the second mesh current and Ohm's law. Compute these values and record them in Table 14.1.

3. Build the circuit of Figure 14.1 using the values specified in step one. Measure the three branch currents and the voltage at node A and record in Table 14.1. Be sure to note the directions and polarities. Finally, determine and record the deviations in Table 14.1.

4. Consider the dual supply circuit of Figure 14.2 using E1 = 10 volts, E2 = 15 volts, R1 = 4.7 k, R2 = 6.8 k, R3 = 10 k, R4 = 22 k and R5 = 33 k. This circuit will require three loops to describe fully. This means that there will be three mesh currents in spite of the fact that there are five branch currents. The three mesh currents correspond to the currents through R1, R2, and R4.

5. Using KVL, write the loop expressions for these loops and then solve to find the mesh currents. Note that the voltages at nodes A and B can be determined using the mesh currents and Ohm's law. Compute these values and record them in Table 14.2.

Laboratory Manual for DC Electrical Circuit Analysis

6. Build the circuit of Figure 14.2 using the values specified in step four. Measure the three mesh currents and the voltages at node A, node B, and from node A to B, and record in Table 14.2. Be sure to note the directions and polarities. Finally, determine and record the deviations in Table 14.2.

Data Tables

Parameter	Theory	Experimental	Deviation
I_{R1}			
I_{R2}			
I_{R3}			
V_A			

Table 14.1

Parameter	Theory	Experimental	Deviation
I_{R1}			
I_{R2}			
I_{R4}			
V_A			
V_B			
V_{AB}			

Table 14.2

Questions

1. Do the polarities of the sources in Figure 14.1 matter as to the resulting currents? Will the magnitudes of the currents be the same if one or both sources have an inverted polarity?

2. In both circuits of this exercise the negative terminals of the sources are connected to ground. Is this a requirement for mesh analysis? What would happen to the mesh currents if the positions of E1 and R1 in Figure 14.1 were swapped?

3. The circuits of Figures 14.1 and 14.2 had been analyzed previously in the superposition theorem and nodal analysis exercises. How do the results of this exercise compare to the earlier results? Should the resulting currents and voltages be identical? If not, what sort of things might affect the outcome?

4. In general, compare and contrast the application of superposition, mesh and nodal analyses to multi-source DC circuits. What are the advantages and disadvantages of each? Are some circuits better approached with a particular technique? Will each technique enable any particular current or voltage to be found or are there limitations?

Laboratory Manual for DC Electrical Circuit Analysis

15

Capacitors and Inductors

Objective

The objective of this exercise is to become familiar with the basic behavior of capacitors and inductors. This includes determination of the equivalent of series and parallel combinations of each, the division of voltage among capacitors in series, and the steady state behavior of simple RLC circuits.

Theory Overview

The inductor behaves identically to the resistor in terms of series and parallel combinations. That is, the equivalent of a series connection of inductors is simply the sum of the values. For a parallel connection of inductors either the product-sum rule or the "reciprocal of the sum of the reciprocals" rule may be used. Capacitors, in contrast, behave in an opposite manner. The equivalent of a parallel grouping of capacitors is simply the sum of the capacitances while a series connection must be treated with the product-sum or reciprocal rules.

For circuit analysis in the steady state case, inductors may be treated as shorts (or for more accuracy, as a small resistance known as the coil resistance, R_{coil}, which is dependent on the construction of the device) while capacitors may be treated as opens. If multiple capacitors are in series, the applied voltage will be split among them inversely to the capacitance. That is, the largest capacitors will drop the smallest voltages.

Equipment

(1) Adjustable DC power supply   model:_____ srn:_____
(1) RLC impedance meter   model:_____ srn:_____
(1) Digital multimeter   model:_____ srn:_____
(1) Electrostatic voltmeter (optional)   model:_____ srn:_____
(1) 4.7 kΩ   _____
(1) 10 kΩ   _____
(1) 100 nF   _____
(1) 220 nF   _____
(1) 1 mH   _____
(1) 10 mH   _____

Schematics

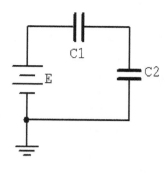
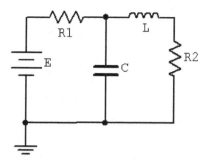

Figure 15.1   Figure 15.2

Procedure

1. Using an RLC meter, measure the values of the two capacitors and two inductors and record them in Table 15.1. Also, measure the equivalent DC series resistance of the two inductors and record them in Table 15.1. Using these values, determine and record the theoretical series and parallel combinations specified in Table 15.2.

2. Connect the two capacitors in series and measure the total capacitance using the RLC meter. Record this value in Table 15.2. Repeat this process for the remaining combinations in Table 15.2. Also determine and record the deviations.

3. Consider the circuit of Figure 15.1 using E = 5 volts, C1 = 100 nF and C2 = 220 nF. Determine the voltage across each capacitor and record these values in Table 15.3.

4. *Only perform this step if an electrostatic voltmeter is available for measurements as a typical DMM will give incorrect results due to loading effects.* Build the circuit of Figure 15.1 using E = 5 volts, C1 = 100 nF and C2 = 220 nF. Measure the voltage across each capacitor and record these values in Table 15.3. Also determine and record the deviations.

5. Consider the circuit of Figure 15.2 using E = 10 volts, R1 = 4.7 k, R2 = 10 k, C = 100 nF and L = 1 mH. Determine the steady state voltage across each component and record these values in Table 15.4.

6. Build the circuit of Figure 15.2 using E = 10 volts, R1 = 4.7 k, R2 = 10 k, C = 100 nF and L = 1 mH. Energize the circuit. It will reach steady state in less than one second. Measure the steady state voltage across each component and record these values in Table 15.4. Also determine and record the deviations.

Data Tables

Series and Parallel Combinations

Component	Experimental	R_{coil}
100 nF		X
220 nF		X
1 mH		
10 mH		

Table 15.1

Pairing	Theory	Experimental	Deviation
100 nF series 220 nF			
100 nF parallel 220 nF			
1 mH series 10 mH			
1 mH parallel 10 mH			

Table 15.2

Capacitive Series Circuit

Voltage	Theory	Experimental	Deviation
V_{C1}			
V_{C2}			

Table 15.3

Steady State RLC Circuit

Voltage	Theory	Experimental	Deviation
V_{R1}			
V_{R2}			
V_C			
V_L			

Table 15.4

Questions

1. Does the value of R_{coil} appear to be correlated with the inductance value?

2. How do capacitors and inductors in series and in parallel compare with resistors?

3. In a series combination of capacitors, how does the voltage divide up?

4. For DC steady state analysis, what can be said about capacitors and inductors?

5. Does the value of R_{coil} seem to have much impact on the final circuit? Why/why not?

Appendix A: Technical Report Guidelines

It is essential that individuals be able to express their ideas and defend their arguments with clarity, detail and subtlety. Similarly, it is important that they can read and critique the ideas and arguments of others in like manner. The creation of lab reports assists in this endeavor. All reports should be neat and legible. Standard technical writing style is expected along with proper grammar and spelling. This means that active voice, first person, personal pronouns, and the like should be avoided. For example, don't write "I set the power supply to 6 volts". Instead use "The power supply was set to 6 volts". **Reports are an individual endeavor.** Although it is perfectly fine to discuss your data and experimental results with your lab partner, the creation of the report itself is an individual exercise. Plagiarism will not be tolerated. A report should conform to the following outline, in the order given:

1. **General Info.** Title, date, your name, partners name.

2. **Objective (AKA Hypothesis).** Answer the question: "What is/are the item(s) under investigation and their proposed relationship(s)?" These are statements of the items that you are testing in this particular exercise.

3. **Conclusion.** Answer the question "What was shown/verified?" These are concise statements of fact regarding the circuit action(s) under investigation. Make sure that you have moved from the *specific lab situation* to the *general case*. If all works well, these should match nicely with your Objective section. Under no circumstances should you reach a conclusion that is not supported by your data, even if that conclusion is stated in the text or in lecture. What matters here is what **you** did and your analysis of it. If there is a discrepancy between your results and theory, state the discrepancy and **don't** ignore your results.

4. **Discussion (AKA Analysis).** Reduce and analyze your data. Explain circuit action or concepts under investigation. Relate theoretical results to the lab results. Don't just state *what* happened, but comment on *why* and its implications. Derive your conclusions from this section. Any deviations from the given procedure (lab manual or handout) must be noted in this section. The Discussion is the penultimate part that you write.

5. **Final Data Sheet.** Include all derived and calculated data. Make sure that you include percent deviations for each theory/measurement pair. Use Percent Deviation = (Measured − Theory) / Theory * 100, and include the sign.

6. **Graphs, Answers to questions at the end of the exercise, Other.** All graphs must be properly titled, created using appropriate scales, and identified with labels. It is suggested that graphs be created with a plotting program or a spreadsheet. Alternately, graphs may be created manually but must be drawn using either a straight edge or a french curve (depending on the type of graph) on appropriate graph paper.

Make sure that you leave sufficient space in the margins and between sections for my comments. Either 1.5 or double line spacing is fine. Multi-page reports must be stapled in the upper left corner. Paper clips, fold-overs, bits of hook-up wire, etc. are not acceptable. Below is the grading standard.

Grade of A: The report meets or exceeds the assignment particulars. The report is neat and professional in appearance, including proper spelling and syntax. The analysis is at the appropriate level and of sufficient detail. Data tables and graphical data are presented in a clear and concise manner. Problem solutions are sufficiently detailed and correct. Diagrams have a professional appearance.

Grade of B: The report is close to the ideal although it suffers from some minor drawbacks which may include some spelling or grammatical errors, analyses which may lack sufficient detail, minor omissions in tabular or graphical data, and the like. In general, the report is solid but could use refinement or tightening.

Grade of C: The report is serviceable and conveys the major ideas although it may be vague in spots. Spelling and grammatical errors may be more numerous than those found in a grade A or B report. Some gaps in data or omissions in explanations may be seen.

Grade of D: Besides typical spelling and grammatical errors, the report suffers from logical errors such as conclusions which are not supported by laboratory data. Analyses tend to be vague and possibly misleading. Graphs and diagrams are drawn in an unclear manner.

Grade of F: The report exhibits many of the following deficiencies: Excessive spelling and grammatical errors, missing sections such as graphs, tables, and analyses, blatantly incorrect analyses, wayward or incomprehensible data, problem solutions tend to be incorrect or missing, and graphical data or diagrams are presented in a shoddy manner.

Appendix B: An Example Technical Report

What follows, starting on the next page, is an example of a technical laboratory report. Read the example *after* reading the report guidelines above. This uses the **non-formal** style.

The experiment in question is completely fabricated, but the report will illustrate both the expected form and content. The mock experiment involves measuring the speed of sound in various materials and whether or not this speed is affected by temperature. In this exercise, the experimenter has affixed small transducers to each end of a solid bar of the material under investigation (rather like a small loudspeaker and microphone). A pulse is then applied to one end and a timer is used to determine how long it takes for the wave to reach the other end. Knowing the length of the bar, the velocity may be computed. The bars are then heated to different temperatures and the process repeated to see if the velocity changes. Appropriate tables and graphs are presented.

The report uses 12 point Times Roman font with 1.5 line spacing although 11 or even 10 point may be preferred. There is no reason to "get fancy" with the appearance of the report. In fact, this will only serve as a distraction. Sufficient space is left for the instructor to insert comments. The length of any specific report can vary greatly depending on the amount of data recorded, the depth of analysis, added graphs, and the like.

As is sometimes the case, this mock experiment didn't work perfectly.

Speed of Sound in Various Materials

Science of Stuff, ET301   February 30, 2112

Name: Johan Bruhaus
Partner: Gail Faux

Objective

The hypothesis investigated in this exercise is straight-forward, namely that the speed of propagation of sound depends on the characteristics of the material and that it may be affected by temperature. Three different materials will be investigated, each at three different temperatures. It is expected that the velocity in all three materials will be significantly greater than the velocity of sound in air (343 meters per second).

Conclusion

The speed of sound in a particular material depends on the internal characteristics of the material. The speed may either increase or decrease with temperature. The velocity at room temperature for the SB alloy was approximately 2001 meters per second with a temperature coefficient (TC) of 0.01%. The GA alloy was 3050 meters per second with −0.2% TC, and the CCCD material was measured at 997 meters per second with 0.1% TC. All values were within a few percent of those predicted by theory, and all velocities were clearly much greater than the velocity of sound in air.

Discussion

To investigate the speed of sound, three bars of material, each one meter long, were obtained. The first was "Sonic Bronze" or SB, an alloy of tin, copper, zinc, and porcupinium. The second material, "Green Aluminum" or GA, is an alloy of aluminum and kryptonite, while the third, CCCD, is commonly known as "Chocolate Chip Cookie Dough".

An acoustical transducer was attached to each end of the bar under investigation. A pulse was applied to one end and a digital timer was used to determine how long it took for the wave to travel down the bar to the pickup transducer. As each bar was one meter long, the velocity in meters per second is simply 1/time delay. The bar was then placed in an industrial oven and the measurement repeated at temperatures of 75°C and 125°C to compare to the nominal room temperature (25°C) results.

The room temperature results agreed strongly with the published data of the three materials. Comparing Table 1.1 to the 25°C column of Table 1.2 showed a deviation no worse than 1.64% (final column, Table 1.2). The variation between materials is approximately 3:1, indicating how strongly the internal characteristics of the material influence the speed of propagation. The CCCD material, being the most plastic, should have the greatest internal frictional losses, and thus, the slowest velocity of the group. This was the case. The inclusion of porcupinium in the SB alloy was responsible for the modest velocity of this material. The waves have to propagate relatively slowly through the porcupinium compared to the GA alloy which is free of this ingredient. The speed of propagation for all materials was significantly faster than the speed of sound through air. Even the slowest of the group, CCCD, exhibited a velocity nearly three times that of air.

The temperature coefficients also showed tight agreement, and appear to be within just a few percent of the established values. Generally, the velocity increases with temperature, although the GA alloy produced the opposite affect. It is assumed that the inclusion of kryptonite in the alloy may be responsible for this. See Graph 1.1 for details.

There was a practical issue involving the CCCD material. The measurements at 25°C and 75°C were satisfactory, however, when the CCCD bar was removed from the 125°C oven it had changed texture and color to a crispy golden brown and produced a strong, pleasing odor. Consequently, one member of the lab group ate approximately 10 centimeters of the bar before the velocity could be measured. To correct for this, the measured time delay was adjusted by a factor of 1.11 as the bar had been reduced to 90% of its original length.

Data

Material	Velocity (m/s)	Temperature Coefficient (% change per degree C)
SB	2000	0.01
GA	3000	−0.21
CCCD	1000	0.105

Table 1.1

Published Theoretical Velocities and TC

Material	Velocity 25°C (m/s)	Velocity 75°C (m/s)	Velocity 125°C (m/s)	Temperature Coefficient	%Deviation at 25°C
SB	2001	2010	2021	0.01	0.05
GA	3050	2750	2440	−0.2	1.64
CCCD	997	1049	1097*	0.1	−0.3

Table 1.2

Experimental Velocities and TC

* See Discussion for explanation

Laboratory Manual for DC Electrical Circuit Analysis

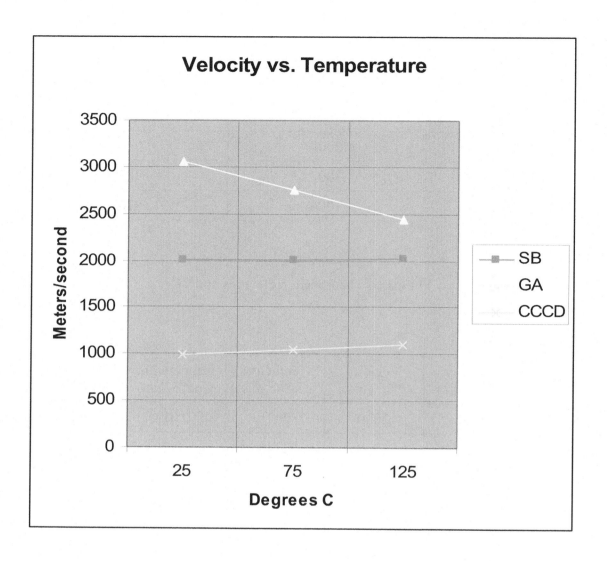

Graph 1.1

Variation of Velocity with Temperature, by Material

Answers to Exercise Questions

1. *Is the velocity of sound unaffected by temperature?*
No. Graph 1.1 shows that in some cases (SB and CCCD) the velocity is directly proportional to temperature although it may be inversely proportional (GA).

2. *If the CCCD material had also been subjected to 175°C, what would you expect?*
It is unlikely that a velocity at 175°C could have been measured as the entire bar probably would have been consumed by the lab team before the transducers could be applied.

Appendix C: Creating Graphs Using a Spreadsheet

While nothing beats good data plotting and analysis software (check out SciDAVis for an excellent free program), you can also create a variety of graphs using spreadsheets such as the one in Open Office or Excel (Microsoft Office). What follows works for Excel 2007 and Open Office 4. Other versions may have different menus and options. Here's how to take your tabular data from lab and create a graph. These instructions assume you will set the independent axis on the horizontal and the dependent axis as the vertical. This is the typical case but there are exceptions (see note at end). Remember, the independent axis presents the input parameter you set (e.g., a power supply voltage or a mass) and the dependent axis presents the output parameter (i.e., the item you are interested in and have measured as an outcome such as a resulting current or change in position).

1. Open a new worksheet. In the first column (column A), enter the text for the legend. This is particularly important if you're plotting multiple datasets on a single graph. Starting in the second column (column B), enter values for the horizontal (independent) axis on the first row of the worksheet. In like fashion, enter values for the vertical (dependent) axis on the second row. For multiple trials, enter the values on subsequent rows. For example, if you are setting a series of voltages in a circuit and then measuring the resulting currents, the voltages would be in row one and the currents in row two. If you changed the circuit components, reset the voltages, remeasured the currents and wish to compare the two trials, then the new set of currents would be in row three and so on. Each of these rows would have their identifying legend in column A with the numeric data starting in column B. Specifically, the legend text for the first data set would be in cell $A$2 and the numeric values would be in cells $B$2 through $X$2 (where X is the final data column), for the second set the legend text would be in cell $A$3 and the numeric values would be in cells $B$3 through $X$3, etc.

2. Select/highlight all of the data (click the first cell, in the upper left corner, and drag the mouse over all of the cells used).

3. Select the Insert menu and choose Chart. Ordinarily you will use an **XY Scatter** chart. There are other options but this is the one you'll need in most cases. A simple Line chart is **not** appropriate in most cases. You might get a graph that "sort of" looks correct but the horizontal axis will simply represent the measurement sequence (first, second, third) rather than the value you set.

4. You can customize the appearance of the chart. In general, you can edit items by simply double-clicking on the item or by using a right-mouse click to bring up a property menu. This will allow you to add or alter gridlines, axes, etc. You can also stipulate variations such as using data smoothing, adding a trend line, etc. It is possible to change the axes to logarithmic or alter their range; and fonts, colors and a variety of secondary characteristics may be altered.

5. Once your chart is completed, you may wish to save the worksheet for future reference. To insert the chart into a lab report, select the chart by clicking on it, copy it to the clipboard (Ctrl+C), select the insertion point in the lab report, and paste (Ctrl+V).

6. In those odd instances where you need to reverse the dependent and independent axes such as a VI plot of a diode where currents are set and resulting voltages are measured, but you want the voltage on the horizontal, some spreadsheets have an axis swap function. If not, you'll need to swap the data ranges for the chart axes. For example, following the instructions above, your independent/horizontal axis is row one. The data are in cells $B$1 through $X$1. The dependent data are in cells $B$2 through $X$2. These ranges can be seen in the chart's Data Series or Data Range menu or dialog box. It will say something like: "X Values: =Sheet1!$B$1:$F$1" and "Y Values: =Sheet1!$B$2:$F$2". Simply swap the row numbers so that it says "X Values: =Sheet1!$B$2:$F$2" and "Y Values: =Sheet1!$B$1:$F$1".

7. Data smoothing can be useful to remove the "jaggyness" of some plots. For simple curves, a second degree B-Spline is suggested if you're using Open Office. For data that are expected to be linear, a trend line can be useful to better see the approximation.

Here is an example worksheet showing a plot of two resistors. The first plot is basic, the second uses smoothed data with a linear trend line:

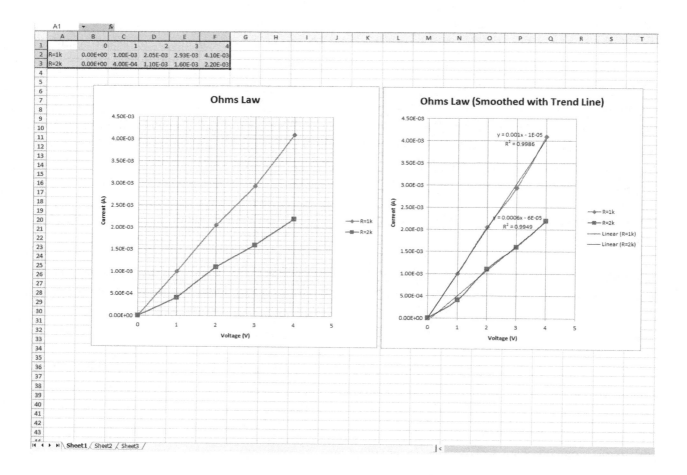

Laboratory Manual for DC Electrical Circuit Analysis

Appendix D: Using a Solderless Breadboard

A solderless prototyping breadboard, or *breadboard* for short, is an essential part of the electrical/electronics toolkit. It allows for the easy assembly, testing and disassembly of all manner of electrical and electronic circuits. Breadboards are available in a wide variety of sizes and shapes, some of which may include a metal mounting plate with binding posts. Whatever the size, all breadboards are configured to provide multiple rows of connecting points using 0.1 inch spacing. This is required when prototyping integrated circuits using through-hole DIP/DIL packages but it also proves to be convenient for small discrete devices such as resistors, diodes, capacitors, transistors and the like. An example is shown in Figure D-1, below.

Figure D-1

This breadboard is shown with an integrated circuit installed. The IC straddles a central trough, its pins inserted into the first hole of each row on either side. All of the five holes in each row are common. For example, directly below the IC is a row highlighted in yellow. These five pins comprise a common wiring point. Similarly, the row highlighted in green presents another set of five common points. Therefore, each pin of the IC has four remaining holes with which to connect to other components. This particular unit contains 63 horizontal rows on either side of the trough, creating 126 sets of five-hole contacts.

Along the left and right edges are double rows of common connection points called *buses*. These are used for tie points that require a large number of connections such as a ground or power supply. This example uses four buses. One of the buses is highlighted in purple. This board has 50 connection points per bus.

To illustrate the internal construction of a breadboard, the bottom cover of one has been removed and the board is shown from the back in Figure D-2. Only a portion of the spring contacts remain, including two long bus contacts shown toward the bottom of the unit. Note how each of the spring contacts is contained in its own isolated "well", effectively insulating it from neighboring rows.

Figure D-2

A close-up of a spring contact is shown in Figure D-3, poised above where it would be positioned. The contact is made of five pairs of small "fingers", each pair corresponding to a connection hole. The metal is fairly thin and thus easily damaged if a wire of excessive diameter is forced in.

Figure D-3

Prototyping a circuit involves translating a schematic onto the breadboard in a clean an efficient manner. The components should not be spaced too far apart nor should they be overcrowded as this will increase the likelihood of accidental shorting of components. Particularly for beginning students, it is often best to echo the overall look of the schematic, placing components in such a way that their location on the breadboard is similar to their location on the schematic. A little forethought regarding circuit layout can also ease the process of parameter measurement, especially when it comes to measuring currents. The process will be illustrated using three examples: a series circuit, a parallel circuit, and a series-parallel circuit.

Series Circuit Example

Consider the schematic shown in Figure D-4. It comprises three resistors connected in series with a single DC voltage source.

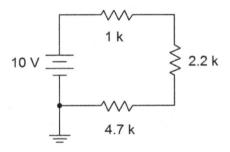

Figure D-4

A common wiring technique is to use the buses for power supply and ground. This is not absolutely necessary for this circuit given that there is only one positive power supply connection and one ground connection, but this a good standard to get used to.

One possible version is shown in Figure D-5. The power supply itself is not shown, however, the red and black connection leads are shown entering from the left side. These connect to the buses via short lengths of solid interconnect wire. The three resistors are then connected in daisy-chain fashion; first from the power bus to a connection row where the second resistor is located. The second resistor jumps to another row of common connection holes, and from there the third resistor jumps down to the ground bus completing the circuit. Note how the layout echoes the original schematic. This helps in the identification of the individual components.

Figure D-5

Of course, there are a great many ways to configure the components on the breadboard. Another possibility is shown in Figure D-6. While this is technically correct, it is not a preferred layout.

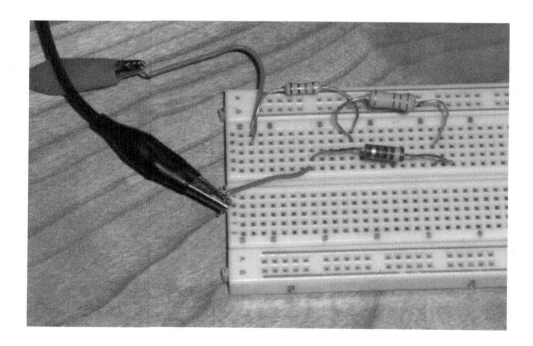

Figure D-6

Laboratory Manual for DC Electrical Circuit Analysis

Generally, voltage measurements are a straight forward affair: simply place the measurement leads across the component(s) to be measured. Figure D-7 shows the connections required to measure the voltage across the 2.2 k ohm resistor. The DMM itself is not included but its leads are shown entering from the right side of the picture.

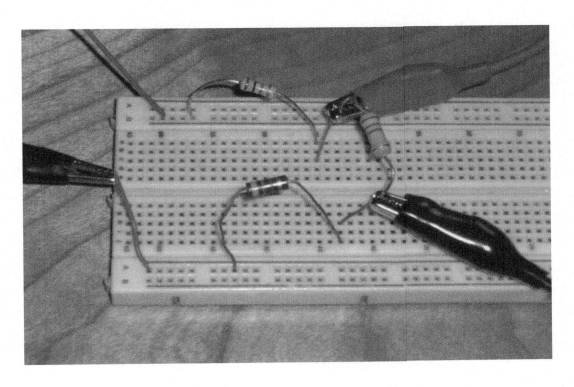

Figure D-7

In contrast, current measurements require that the ammeter be inserted in-line. This means that the circuit must be "broken open" in order to insert the ammeter. To do so, simply move one end of the component of interest into an unused connection row and then connect the ammeter from this point to the original location. This is shown in Figure D-8, showing the measurement of the current flowing through the 2.2 k ohm resistor. The 2.2 k ohm has been moved over a couple of rows and the ammeter is then connected from the original point to this new point. Again, the ammeter itself is not shown although its leads are shown entering from the right.

Finally, note the polarity used for the meter on both the voltage and current measurements. The red lead is placed at the expected positive point (i.e., the more positive voltage of the two points, or in the case of current, the entering point for conventional current flow). Remember, conventional current flows from positive to negative so the red lead should be positive and the black lead at negative. Failure to follow this standard will create ambiguous positive/negative readings.

Figure D-8

Parallel Circuit Example

Consider the schematic shown in Figure D-9. It comprises three resistors connected in parallel with a single DC voltage source.

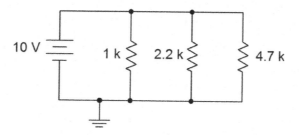

Figure D-9

A parallel circuit can make excellent use of the buses. This is shown in Figure D-10 (again, the power supply leads enter from the left).

Figure D-10

Once again, measurement of the current through a single component requires a slight rearrangement. For example, measuring the current through the 4.7 k ohm resistor requires that the ammeter be inserted between the power bus and the resistor. Thus, the resistor must be moved off of the bus and onto a five hole connection row. The ammeter will then be connected between them, as shown in Figure D-11.

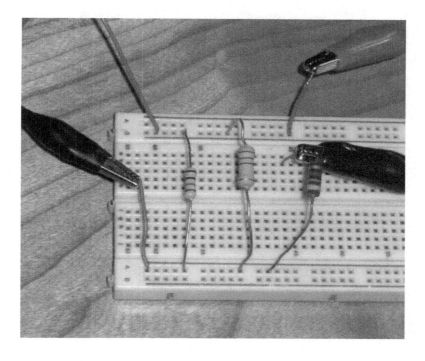

Figure D-11

Laboratory Manual for DC Electrical Circuit Analysis

Series-Parallel Circuit Example

Consider the schematic shown in Figure D-12. It comprises three resistors connected in series-parallel with a single DC voltage source.

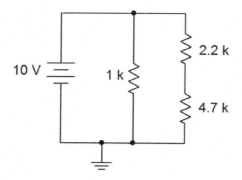

Figure D-12

One possible interconnection is shown in Figure D-13. Note the need to jump the 2.2 k and 4.7 k ohm resistors to an unused common row somewhere on the board.

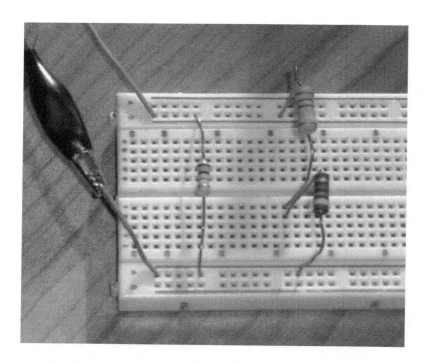

Figure D-13

Once again, voltage measurement is straight forward. The connections to measure the voltage across the 4.7 k ohm resistor are shown in Figure D-14 (voltmeter connection leads on the right side).

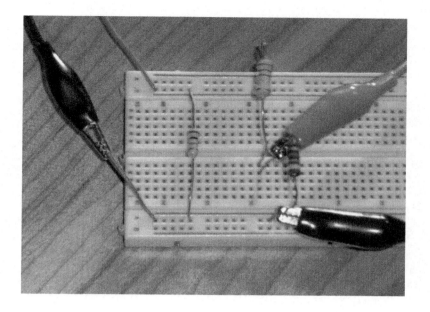

Figure D-14

Current measurement through the 2.2 k ohm resistor is shown in Figure D-15. Note the realignment of the resistor and the ammeter insertion.

Figure D-15

Finally, measurement of the total circuit current (i.e., the supply current) requires inserting the ammeter prior to the power bus. This is shown in Figure D-16. Note that the ammeter's positive lead is connected directly to the power supply lead while the ammeter's black lead is connected back to the power bus (i.e., where the power used to be connected). Thus, all of the current exiting the supply must flow through the ammeter before reaching the circuit, and therefore the ammeter measures the total circuit current.

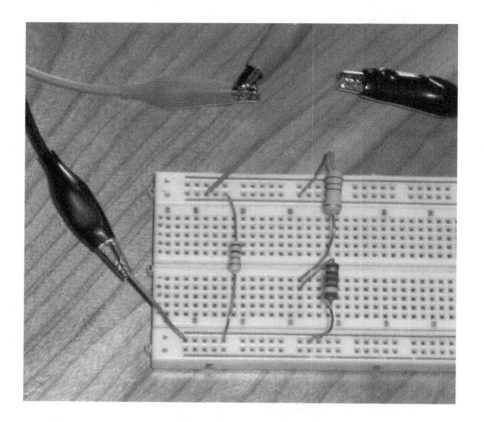

Figure D-16